Cambridge Elements ☰

Elements in Geochemical Tracers in Earth System Science
edited by
Timothy Lyons
University of California
Alexandra Turchyn
University of Cambridge
Chris Reinhard
Georgia Institute of Technology

CERIUM ANOMALIES AND PALEOREDOX

Rosalie Tostevin
University of Cape Town

CAMBRIDGE
UNIVERSITY PRESS

University Printing House, Cambridge CB2 8BS, United Kingdom

One Liberty Plaza, 20th Floor, New York, NY 10006, USA

477 Williamstown Road, Port Melbourne, VIC 3207, Australia

314–321, 3rd Floor, Plot 3, Splendor Forum, Jasola District Centre,
New Delhi – 110025, India

79 Anson Road, #06–04/06, Singapore 079906

Cambridge University Press is part of the University of Cambridge.

It furthers the University's mission by disseminating knowledge in the pursuit of
education, learning, and research at the highest international levels of excellence.

www.cambridge.org
Information on this title: www.cambridge.org/9781108810739
DOI: 10.1017/9781108847223

First published 2021

A catalogue record for this publication is available from the British Library.

ISBN 978-1-108-81073-9 Paperback
ISSN 2515-7027 (online)
ISSN 2515-6454 (print)

Cerium Anomalies and Paleoredox

Elements in Geochemical Tracers in Earth System Science

DOI: 10.1017/9781108847223
First published online: January 2021

Rosalie Tostevin
University of Cape Town
Author for correspondence: Rosalie Tostevin, rosalie.tostevin@uct.ac.za

Abstract: Cerium (Ce) anomalies track changes in oxygen availability due to the anomalous redox sensitivity of Ce compared with the other rare earth elements. The proxy systematics have been calibrated experimentally as well as in modern anoxic water bodies. Ce anomalies are unique because they track intermediate manganous conditions rather than fully anoxic conditions. In addition, they are sensitive to local-regional redox conditions and can be analysed in chemical sediments such as carbonate rocks. This makes them especially useful as a tool to track local oxygen distribution in shallow shelf environments, where biodiversity is highest. This review focusses on the systematics of the Ce anomaly proxy, the preservation and extraction of the signal in sedimentary rocks and the potential applications of the proxy.

Keywords: cerium, oxygen, sedimentary rocks, redox conditions, manganous, geochemical proxies

ISBNs: 9781108810739 (PB), 9781108847223 (OC)
ISSNs: 2515-7027 (online), 2515-6454 (print)

Contents

1 Introduction

The rare earth elements (REE) have very similar chemical properties but cerium (Ce) has unique redox chemistry. In oxygenated waters, Ce^{3+} is oxidised to insoluble Ce^{4+} and accumulates as discrete Ce oxide particles or on the surface of Mn(IV)-(oxyhydr)oxide minerals and other particulate matter. Under intermediate manganous conditions, Ce oxides and Ce-enriched Mn(IV)-oxy(hydr) oxide particles undergo reductive dissolution and excess Ce is released back into the water column. In general, oxygenated modern marine settings display a strong negative Ce anomaly while manganous and anoxic waters lack a negative Ce anomaly (Bau et al., 1997; De Baar et al., 1988; De Carlo and Green, 2002; German et al., 1991). Enrichments or depletions in Ce compared with other REE have been used to track redox cycling in rivers, lakes and oceans over several decades (De Baar et al., 1988; German and Elderfield, 1990a; German et al., 1991). In addition, Ce anomalies can be faithfully preserved in chemical sediments such as phosphorites and carbonates and so provide a potentially useful redox proxy in deep time.

Three characteristics make cerium anomalies particularly useful as a redox proxy. Firstly, the REE provide a robust baseline from which to quantify enrichments or depletions in cerium, as they otherwise have very similar chemical properties and should behave in a predictable, coherent manner. This is in contrast to many other redox-sensitive elements such as U or Mo, which are normalised to major elements (e.g., Ca, Al) that may vary independently. In addition, other features of the REE pattern (e.g., Y/Ho ratios) can be used to confirm the preservation and extraction of pristine seawater signals from ancient rocks. Secondly, Ce is sensitive to local or regional redox conditions unlike many other redox-sensitive elements and isotope systems which respond to the global area of anoxic seafloor. Finally, the reduction potential of Ce(IV) (+1.61˚V) is higher than Mn(IV) (1.23˚V) and its oxidation commonly occurs on the surface of Mn(IV)-(oxyhydr)oxide minerals. Therefore, the redox cycling of Ce is closely related to the redox cycling of Mn and is sensitive to intermediate manganous conditions (often referred to as suboxic), which may overlap with low levels of oxygen. In contrast, the majority of redox proxies respond to fully anoxic conditions (i.e., the onset of ferruginous conditions or euxinia).

Seawater REE patterns, including any Ce anomaly, can be faithfully preserved in a variety of chemical sediments including carbonate rocks (Nothdurft et al., 2004; Webb and Kamber, 2000). Ce anomalies are relatively robust to diagenesis and even low-grade metamorphism and primary signals can often be extracted from bulk rock limestone and even dolomite (Banner et al., 1988;

Hood et al., 2018; Liu et al., 2019; Webb and Kamber, 2000; Webb et al., 2009). Consequently, Ce anomalies have proved popular in paleo-redox studies, with consistent mention in the literature since the 1950s and an increase in attention in the early 2000s (Figure 1). Since carbonate rocks commonly form in shelf environments, Ce anomalies capture information about oxygen availability in shallow waters, where the majority of biodiversity is located. Ce anomalies are therefore particularly useful for understanding the relationship between oxygen availability and animal ecosystems in deep time.

2 Underpinning of the Proxy

Rare earth elements are constantly undergoing exchange equilibrium between REE^{3+} solution complexes (mostly mono- and di-carbonate complexes and minor siderophore, silicate or sulfate complexes) and hydroxide complexes such as $REE(OH)^{2+}$ on the surface of Fe-Mn oxides and hydroxides, clays or organic matter. In oxygenated waters, Ce^{3+} is partially oxidised to Ce^{4+} either abiotically (Bau, 1999; Koeppenkastrop and De Carlo, 1992) or via microbial mediation

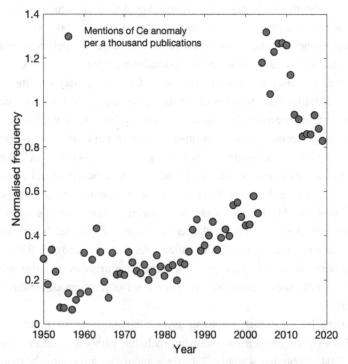

Figure 1 Number of mentions of Ce anomalies in the literature over the past 70 years, normalised to the total number of publications registered on ScienceDirect

(Moffett, 1990). Although discrete Ce(IV)-oxide particles can form, oxidation commonly occurs on the surface of metal oxide minerals. Oxidised Ce^{4+} is insoluble and ceases to participate in exchange reactions with dissolved Ce^{3+}, causing Ce to gradually accumulate on the solid surface over time (Figure 2). This results in a positive Ce anomaly in metal oxides and a corresponding negative Ce anomaly in oxic waters (Sholkovitz et al., 1994). Metal oxides are either buried intact in sediments below oxic bottom waters or they encounter manganous or anoxic waters and undergo reductive dissolution, releasing excess Ce (Figure 2). Therefore, waters beneath a Mn(IV)/Mn(II) redoxcline commonly exhibit either no Ce anomaly or a positive Ce anomaly.

Hydrogenetic crusts and nodules that precipitate slowly from seawater yield positive Ce anomalies and an inverse REE pattern to seawater, supporting strong fractionation during REE accumulation on the surface (Bau et al., 2014). Ce accumulation is associated with Mn(IV)-(oxyhydr)oxides, whereas REE are thought to be scavenged quantitatively onto the surface of Fe(III)-

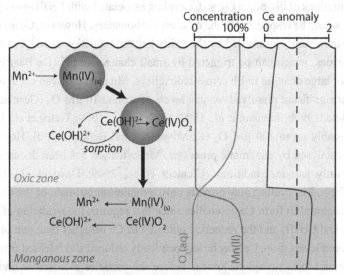

Figure 2 Aqueous $REE(OH)^{2+}$ complexes including $Ce(OH)^{2+}$ undergo exchange reactions on the surface of Mn(IV)-(oxyhydr)oxide minerals (grey circles), which form in the oxic zone of the water column. Ce^{3+} is then oxidised to $Ce(IV)O_2$ and remains on the surface until the Mn(IV)-(oxyhydr)oxide minerals sink beneath the Mn redoxcline and undergo reductive dissolution, releasing any bound Ce. The relative concentrations of oxygen (blue) and Mn(II) (pink) are shown alongside the magnitude of the Ce anomaly (grey). For interpretation of the colours in this figure, the reader is referred to the web version of this Element.

(oxyhydr)oxides (Bau, 1999; De Carlo, 2000; Ohta and Kawabe, 2001). This is supported by evidence from hydrothermal plumes (Edmonds and German, 2004), the water column (De Baar et al., 1988; De Carlo and Green, 2002; German and Elderfield, 1990b; Moffett, 1990; Sholkovitz et al., 1994), marine Fe-Mn nodules (De Carlo, 2000), REE data from banded iron formation (Planavsky et al., 2010), and experimental work (Bau, 1999; Koeppenkastrop and De Carlo, 1992; Ohta and Kawabe, 2001; Quinn et al., 2006). For example, where Mn-oxides have been documented forming *in situ*, they are associated with positive Ce anomalies (Sholkovitz et al., 1994). In contrast, Ce enrichments are absent from Mn-poor hydrothermal nodules (Edmonds and German, 2004). However, sequential leaching of some seafloor nodules produces conflicting results, suggesting Fe oxides play a more important role in the fractionation of rare earth elements and yttrium (REY) during scavenging (Bau and Koschinsky, 2009).

This raises the question of exactly what range of oxygen concentrations is required to generate Ce anomalies. Fe(II) is rapidly oxidised at very low (nM) concentrations of dissolved O_2, so Ce cycling associated with Fe(III)-(oxyhydr) oxides would be responsive to the oxic/anoxic boundary. However, Ce anomalies are dominantly controlled by the oxidation and reduction of Mn(IV)-(oxyhydr)oxides, which can be triggered by small changes in pH (De Baar et al., 1988) or large changes in Eh across redoxclines. Mn reduction can occur under low but significant dissolved oxygen levels of around 10 μM O_2 (German and Elderfield, 1990b; Johnson et al., 1992; Saager et al., 1989; Trefry et al., 1984) and possibly up to 100 μM O_2 (Klinkhammer and Bender, 1980). However, where catalysed by enzymatic processes, Mn oxidation has been documented under fully anoxic conditions (Clement et al., 2009; Daye et al., 2019). Regardless, Mn cycling may provide only a minimum estimate of the oxygen levels required to form Ce anomalies since this requires the oxidation of both Mn(II) and Ce(III) and the reduction potential of Ce is higher than that of Mn. As oxygen levels drop, Ce may be independently reduced and released from the surface of Mn(IV)-(oxyhydr)oxides before the particles themselves undergo reductive dissolution. There is even evidence to suggest that Ce may form discrete CeO_2 particles, decoupling Ce cycling from Mn cycling altogether (Haley et al., 2004). Therefore, a positive Ce anomaly or the absence of a Ce anomaly indicates that a sample is from below the Mn redoxcline but does not distinguish between manganous or fully anoxic waters.

Ce anomaly data are most powerful when combined with additional redox constraints. For example, Fe speciation is responsive to the onset of anoxic conditions and when combined with Ce anomaly data can provide a unique marker for intermediate manganous conditions (Clarkson et al., 2014; Tostevin

The natural concentration of REE when plotted against atomic number forms a zig-zag pattern with enrichments in even-numbered REE. The REE concentrations are therefore normalised to shale (post-Archean average shale (PAAS); Pourmand et al., 2012), and plotted on a log-scale to highlight meaningful trends and anomalies. Shale is used because it has a similar composition to the upper continental crust, the main source of REE to seawater. Ce anomalies (Ce/Ce*) are calculated based on relative enrichments or depletions in shale-normalised Ce ($[Ce]_{SN}$) compared to neighbouring non-redox-sensitive REY (Lawrence and Kamber, 2006). Historically, Ce anomalies have been calculated based on La and Pr concentrations:

$$Ce/Ce^* = \frac{[Ce]_{SN}}{0.5[La]_{SN} + 0.5[Pr]_{SN}}$$

La can also show anomalous enrichments in seawater, so Lawrence and colleagues (2006) proposed a new calculation that avoids comparison with La:

$$Ce/Ce^* = \frac{[Ce]_{SN}}{([Pr]_{SN})^2 / [Nd]_{SN}}$$

The cut-off value used to define Ce anomalies is arbitrary but many studies use a conservative range of <0.8 for negative anomalies and >1.2 for positive anomalies.

Seawater REE and associated Ce anomalies can be faithfully preserved in ironstones, phosphate minerals and carbonate rocks without fractionation under the right conditions (Planavsky et al., 2010; Shields and Stille, 2001; Webb and Kamber, 2000). Fe(III)-Mn(IV) (oxyhydr)oxide minerals that precipitate rapidly as hot hydrothermal fluids mix with seawater yield a seawater REE pattern, suggesting they could be used as a proxy archive for paleo-seawater (Bau et al., 2014; Planavsky et al., 2010). However, even after exchange reactions have ceased for the other REE, Ce can continue to accumulate on the surface (Bau and Koschinsky, 2009).

Seawater REE can also be preserved in phosphate minerals under some conditions. This is supported by examples of covariation between Ce anomalies and other primary redox indicators such as bioturbation (Macleod and Irving, 1996; Shields and Stille, 2001). However, phosphate minerals scavenge REE non-quantitatively and are highly susceptible to overprinting due to continued growth during early diagenesis (Shields and Stille, 2001; Shields and Webb, 2004). This typically leads to non-seawater REE patterns with enrichments in the middle REE and a weakened Ce anomaly. Granular phosphorites are less susceptible to alteration than skeletal phosphate because they undergo extensive reworking at the seafloor in contact with seawater but their formation is biased towards upwelling margins and the Phanerozoic eon. Carbonate minerals offer a more reliable seawater REE archive and are widespread throughout geological time.

Rare earth elements are enriched in carbonate rocks compared with solution but the partition coefficient is similar for all REE. In general, non-skeletal archives, including stromatolites, ooids and marine cements, are considered more reliable, which is useful as they are common throughout the rock record (Li et al., 2019; Tostevin et al., 2016a; Webb and Kamber, 2000). Some simple skeletal organisms, such as scleractinian corals, preserve seawater REE (Webb et al., 2009). However, more complex skeletal carbonates may be impacted by metabolic processes or association with organic material and metal oxide coatings. For example, some brachiopod species record altered REE patterns in the primary layer and umbo region but primary signals in the dorsal and ventral valves (Zaky et al., 2016). Similarly, bivalves shells record a non-seawater REE signal because they reflect the mixture of seawater, food and other particulate matter that enters through the siphon (Akagi and Edanami, 2017). Pristine foraminifera may record primary seawater REE but specialised cleaning procedures must be used to remove organic and oxide coatings (Haley et al., 2005).

Diagenetic alteration is a common concern in carbonate rocks because unstable primary minerals such as aragonite and high-Mg calcite can recrystallise to more stable low-Mg calcite during burial. This neomorphism commonly occurs in the presence of pore fluids with radically different redox conditions and chemical compositions compared with overlying seawater. The REE composition of pore fluids is distinct from seawater and evolves with depth as organic carbon, phosphate minerals and Fe-Mn (oxyhydr)oxides dissolve in sequence, releasing REE (Haley et al., 2004; Kim et al., 2012). In addition, the absence of positive Ce anomalies in diagenetic metal oxides suggests Ce is not readily mobilised during metal cycling in pore waters (Bau et al., 2014). Despite this, carbonate-bound REE appear relatively robust to diagenetic alteration (Webb and Kamber, 2000). Since REE are structurally incorporated, replacing the calcium ion, there is a low risk of exchange as water flows through the rocks. In addition, REE/Ca ratios in diagenetic fluids are typically low. Primary REE can be preserved during meteoric and anoxic marine burial diagenesis even where other trace element systems are compromised (Azmy et al., 2011; Hood et al., 2018; Liu et al., 2019; Nothdurft et al., 2004). Where alteration does occur, during extensive open-system diagenesis in the presence of fluids with high concentrations of non-seawater REE, the major features of the REE pattern are retained (Azmy et al., 2011; Hood et al., 2018; Nothdurft et al., 2004). The REE patterns in carbonate rocks that have similar depositional histories but have experienced different diagenetic conditions and episodes of dolomitisation suggest primary REE patterns can be retained in dolomite (Banner et al., 1988; Hood et al., 2018; Liu et al., 2019).

A bulk carbonate rock may contain several REE-bearing phases such as limestone, dolomite, phosphate, oxides, clays and organic matter. Each of these phases will be associated with a unique REE signature. It is important to isolate the carbonate signal during REE extraction to obtain a pristine seawater signal. Clays, for example, contain very high concentrations of REE with a flat continental pattern, and even small amounts of clay leaching (<1%) can overwhelm the carbonate signature (Nothdurft et al., 2004). Sequential leaching techniques enable the carbonate phase to be isolated, preventing contributions from oxides, phosphates or clay minerals (Cao et al., 2020; Tostevin et al., 2016a; Zhang et al., 2015). Recent work using experimental mixtures of different phases demonstrated that acetic acid is preferable to nitric acid and highlighted the importance of filtering the supernate (Cao et al., 2020). Rare earth elements can be extracted from dolomite-rich rocks but a modified leaching procedure is required (Tostevin et al., 2016a). Even with optimised leaching procedures, samples with <85% $CaCO_3$ are unlikely to be successful (Cao et al., 2020; Tostevin et al., 2016a). Additional information could be gained by intentionally targeting non-carbonate phases to isolate REE signals, but regardless of the aims of the study it is important to isolate each phase to avoid a mixed signal.

Geochemical indicators such as Mn/Sr, Sr/Ca and $\delta^{18}O$ are commonly used to screen carbonate rocks for contamination or alteration. Given that Ce anomalies are considered more robust to alteration than these indicators, an unaltered signal provides reassurance that the Ce anomaly is also preserved. In cases where these indicators have been altered they are less useful because it is not clear whether the Ce anomaly has also been compromised. The fidelity of the Ce anomaly is best assessed by studying the REE pattern itself because particular features, such as enrichments in the heavy REE, Y/Ho ratios >36, small positive Eu anomalies and small positive La anomalies, indicate preservation of a seawater signal. Any alteration of the REE pattern should also impact the Ce anomaly unless there is independent mobilisation of Ce during redox cycling. Geochemical data can also be used to detect unintended leaching of non-target phases. For example, any Al in the leachate is likely derived from the partial dissolution of clay minerals. Similarly, Y/Ho ratios are lower in leachates from samples with a higher proportion of oxide or clay minerals, indicating the increased probability of contributions from these phases (Tostevin et al., 2016a). Leaching of non-target phases also increases the total REE content of the leachate ($\sum$REE). As such, threshold values of Al < 1 ppm, $\sum$REE < 1 ppm and Y/Ho > 36 are commonly used to screen REE data. However, Y/Ho ratios also vary naturally in seawater, particularly across redox boundaries, as do primary REE contents in various carbonate phases, so cut-off values must be considered on a case-by-case basis. Geochemical cut-off values offer an easy way to screen large data sets but the threshold values are arbitrary and care must be taken to use multiple screening techniques including petrography.

4 Case Studies

4.1 Modern Validation

The Ce anomaly proxy has been calibrated in modern settings including inland seas and lakes such as the Black Sea and Lake Vanda, and in restricted marine environments such as the eastern Mediterranean and the Cariaco Basin (Figure 3) (Bau et al., 1997; De Baar et al., 1988; De Carlo and Green, 2002). Early water column data demonstrated that negative Ce anomalies develop in oxic surface waters and are rapidly eroded in anoxic deep waters, coincident with increases in dissolved Mn^{2+} and Fe^{2+}. For example, oxic waters of the Tyro Sub-basin in the eastern Mediterranean yield a negative Ce anomaly at 3,200 m depth (Bau et al., 1997). The next available REE data, at 3,400 m depth, come from anoxic hypersaline brines and yield a small positive Ce anomaly. The oxic/anoxic boundary is located within this interval, between 3,370 and 3,390 m depth (Saager et al., 1993). Similarly, Ce anomalies in the Cariaco Trench, a 1,400 m deep depression in the Caribbean Sea, rapidly transition from a large negative to a small positive anomaly between 290 and 295 m depth (De Baar et al., 1988), coincident with an increase in dissolved Mn^{2+}, Fe^{2+} and H_2S (Jacobs et al., 1987). In both the eastern Mediterranean and the Cariaco Trench, the manganous zone, if present, is too thin to detect at the available sample resolution.

More detailed data from the Black Sea, the Saanich Inlet and Lake Vanda revealed that Ce cycling occurs under intermediate manganous conditions before the onset of anoxia. The Black Sea is a large, stable basin with permanently anoxic deep waters. The Ce anomaly increases with depth, reaching a minima (0.05) at 76 m, coincident with a maxima in dissolved particulate Mn concentrations. Below this, the Ce anomaly increases dramatically to ~1. Dissolved Fe^{2+} concentrations don't begin to increase until 110 m, coincident with the onset of fully anoxic conditions. This implies the existence of a significant intermediate zone and suggests Ce anomalies are responsive to the onset of manganous rather than ferruginous conditions. A similar pattern was observed in the seasonally anoxic Saanich Inlet basin in British Columbia (German and Elderfield, 1989), where redox cycling of Ce and Mn occurs in a distinct manganous zone 20 m above the onset of anoxia.

The relationship between Ce and Mn is particularly clear in Lake Vanda, Antarctica, an inversely stratified saline lake with warm, saline anoxic bottom waters overlain by cool, oxygenated freshwater. The permanent ice cover prevents wind-

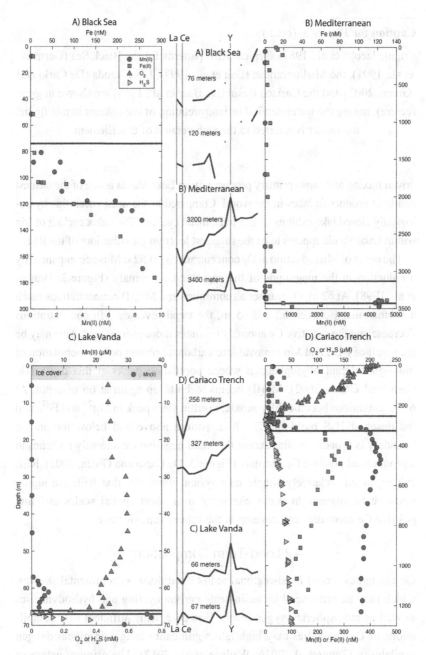

Figure 3 Water column data including dissolved O_2 (blue upward-facing triangles), H_2S (yellow right-facing triangles), Fe(II) (green squares) and Mn(II) (pink circles) from A) the Black Sea (top left; Lewis and Landing, 1991); B) the eastern Mediterranean (top right; Saager et al., 1993); C) Lake Vanda, Antarctica (bottom left; Bratina et al., 1998); and D) the Cariaco Trench (bottom

---

Caption for Figure 3 (cont.)

right; Jacobs et al., 1987). Typical REE patterns for the Black Sea (German et al., 1991), the Mediterranean (Bau et al., 1997), Lake Vanda (De Carlo and Green, 2002) and the Cariaco Basin (De Baar et al., 1988) are shown in grey (centre), noting the water depth. For interpretation of the colours in this figure, the reader is referred to the web version of this Element.

driven mixing and limits primary productivity so Lake Vanda is one of the clearest and least productive lakes in the world. Compared to the open ocean this hydrologically closed lake exhibits simple trace metal cycling. The redox cycling of Mn within Lake Vanda appears to be the strongest lever on the behaviour of the REE.

The onset of Mn reduction at O_2 concentrations <0.52 mM is accompanied by a reduction in the magnitude of the negative Ce anomaly (Figure 3; Bratina et al., 1998). At 62 m, O_2 reaches a subminima and Mn(II) concentrations reach a submaxima. Between 62 and 65 m, the trend reverses: Mn concentrations decrease and the negative Ce anomaly becomes more pronounced. This may be the result of REE and Mn removal into carbonate phases or by a population of low-light-tolerant phytoplankton whose population peaks at this depth (De Carlo and Green, 2002). Mn(II) begins to build up again at 66 m depth. At 67 m, conditions become fully anoxic, resulting in a peak in Mn^{2+} and Fe^{2+} and the onset of H_2S build-up. The REE pattern above and below the anoxic boundary is broadly similar except the small negative Ce anomaly switches to a pronounced positive Ce anomaly (Figure 3; De Carlo and Green, 2002). REE cycling in this relatively simple lake system illustrates that REE are highly sensitive to changes in water chemistry over short spatial scales and that positive Ce anomalies can develop in the suboxic-anoxic zone.

4.2 Long-Term Compilations

Ce anomalies record local-regional redox conditions on continental shelves, which may be influenced by basin-scale carbon cycling and hydrodynamics as well as atmospheric oxygen levels. However, with sufficient data compilations of Ce anomalies may highlight significant changes in global oxygen availability (Tang et al., 2016; Wallace et al., 2017). The primary nature of the negative Ce anomalies recorded at 3.2 Ga (Kato et al., 1998) and 2.8 Ga (Kato et al., 2002) has been questioned because these studies didn't measure Pr and the anomalies occur randomly with respect to stratigraphy (Planavsky et al., 2010). Aside from these potentially altered samples, ironstones and carbonate rocks record broadly consistent signals throughout the Archean and Proterozoic, suggesting the oceans were dominated by anoxic waters. Surface waters would have contained low concentrations of $O_2(aq)$ throughout the Proterozoic but the concentration and depth of the oxic layer was clearly limited. A significant change in the record is evident in the Phanerozoic (Figure 4), suggesting widespread, well-oxygenated waters became a persistent feature of shelf environments from the Devonian onwards (Figure 4) (Tostevin and Mills, 2020; Wallace et al., 2017).

5 Future Prospects

Our understanding of Ce anomalies has evolved through decades of research. The basic principle – that the unique redox-sensitive behaviour of Ce results in depletions or enrichments compared to neighbouring REE – is well established (German and Elderfield, 1990a; German et al., 1991). However, more detailed work is needed to understand the kinetics of Ce oxidation, the impact of organic matter and the dependence of Ce oxidation on Mn particulates (Kim et al., 2012).

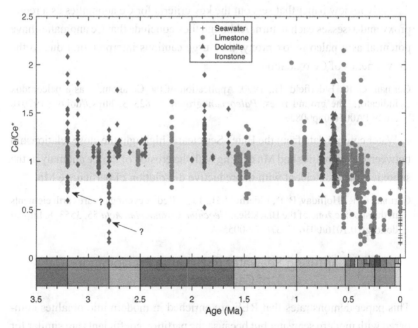

Figure 4 Ce anomaly data through geological time, building on data sets compiled by Wallace et al. (2017) and Tang et al. (2016). The grey shaded region indicates no significant Ce anomaly (0.8–1.2). Arrows highlight negative anomalies that have been questioned by Planavsky et al. (2010). Geological eras and periods are marked along the axis (refer to GSA Geological Timescale v. 5.0 for interpretation of colours).

A large body of work has demonstrated that seawater signals can be well preserved in carbonate rocks and sediments (Nothdurft et al., 2004; Webb and Kamber, 2000) and that with careful leaching procedures they can be successfully extracted without contamination from non-carbonate phases (Cao et al., 2020; Tostevin et al., 2016a; Zhang et al., 2015). The preservation of seawater signals in bulk limestone and dolomite means the proxy is particularly relevant to Precambrian rocks deposited before the evolution of carbonate skeletons. Ce anomalies offer a promising pathway for tracking redox conditions in deep time. The relative ease of preparation and analysis means large data sets can be collected that map variations in redox conditions across a single shelf environment (Tostevin et al., 2016b). Ce anomalies may provide particularly useful insight into the role of oxygen in early evolution since they are sensitive to intermediate conditions that are more relevant to animal ecosystems.

6 Key Papers
Establishing the Redox-Sensitive Behaviour of Ce

An early review paper that sets out the key criteria for Ce anomalies as a redox proxy and assesses each in turn. Although they conclude that Ce anomalies have potential as a paleo-redox proxy, they urge cautious interpretation due to the slow kinetics of Ce oxidation.

German, C. R., Elderfield, H., 1990. Application of the Ce anomaly as a paleoredox indicator: The ground rules. *Paleoceanography* 5, 823–3. https://doi.org/199010 .1029/PA005i005p00823

Water column data from the Black Sea clearly highlight the close relationship between Ce anomalies and Mn cycling with the erosion of the Ce anomaly in the suboxic zone, coincident with the reductive dissolution of particulate Mn.

German, C. R., Holliday, B. P., Elderfield, H., 1991. Redox cycling of rare earth elements in the suboxic zone of the Black Sea. *Geochim. Cosmochim. Acta* 55, 3553–8. https:// doi.org/10.1016/0016–7037(91)90055-A

Capturing Seawater REE during Precipitation of Carbonate Rocks

This paper demonstrates that REE are enriched in modern microbialites compared with modern seawater but because the partition coefficients are similar for all REE, patterns and anomalies are faithfully preserved.

Webb, G. E., Kamber, B. S., 2000. Rare earth elements in Holocene reefal microbialites: A new shallow seawater proxy. *Geochim. Cosmochim. Acta* 64, 1557–65. https://doi .org/10.1016/S0016-7037(99)00400–7

This study tested the application of REE in ancient carbonate rocks by comparing REE patterns in various carbonate phases precipitated from the same body of seawater. They found small differences in REE patterns due to variations in partitioning behaviour but conclude that, overall, carbonate rocks have potential as ancient seawater REE archives.

Nothdurft, L. D., Webb, G. E., Kamber, B. S., 2004. Rare earth element geochemistry of Late Devonian reefal carbonates, Canning Basin, Western Australia: Confirmation of a seawater REE proxy in ancient limestones. *Geochim. Cosmochim. Acta* 68, 263–83. https://doi.org/10.1016/S0016-7037(03)00422-8

Preservation and Extraction of REE Signals in Carbonate Rocks

This study demonstrates the high preservation potential of Ce anomalies in recent carbonate sediments that have experienced a range of diagenetic conditions including meteoric processes and marine burial diagenesis, including dolomitisation.

Liu, X.-M., Hardisty, D. S., Lyons, T. W., Swart, P. K., 2019. Evaluating the fidelity of the cerium paleoredox tracer during variable carbonate diagenesis on the Great Bahamas Bank. *Geochim. Cosmochim. Acta* 248, 25–42. https://doi.org/10.1016/j.gca.2018.12.028

This early work demonstrated that seawater REE patterns can be extracted from carbonate rocks that have undergone episodes of regional dolomitisation.

Banner, J. L., Hanson, G. N., Meyers, W. J., 1988. Rare earth element and Nd isotopic variations in regionally extensive dolomites from the Burlington-Keokuk Formation (Mississippian): Implications for REE mobility during carbonate diagenesis. *J. Sediment. Res.* 58, 415–32.

This paper demonstrates the impact of different leaching procedures on the REE pattern extracted from carbonate rocks and sediments, showing how easily the seawater signal can be overprinted by contamination.

Tostevin, R., Shields, G. A., Tarbuck, G. M., He, T., Clarkson, M. O., Wood, R.A., 2016. Effective use of cerium anomalies as a redox proxy in carbonate-dominated marine settings. *Chem. Geol.* 438, 146–62. https://doi.org/10.1016/j.chemgeo.2016.06.027

Application in Deep Time

This study compiles Ce anomaly data from the Neoproterozoic and Palaeozoic and identifies a significant long-term shift in the average. This work is significant because it shows that with sufficient data global long-term trends can be extracted from local proxy data.

Wallace, M. W., Hood, A. vS., Shuster, A., Greig, A., Planavsky, N. J., Reed, C. P., 2017. Oxygenation history of the Neoproterozoic to early Phanerozoic and the rise of land plants. *Earth Planet. Sci. Lett.* 466, 12–19. https://doi.org/10.1016/j.epsl.2017.02.046

References

Akagi, T., Edanami, K., 2017. Sources of rare earth elements in shells and soft tissues of bivalves from Tokyo Bay. *Mar. Chem.* 194, 55–62. https://doi.org /10.1016/j.marchem.2017.02.009

Alibo, D. S., Nozaki, Y., 1999. Rare earth elements in seawater: Particle association, shale-normalization, and Ce oxidation. *Geochim. Cosmochim. Acta* 63, 363–72. https://doi.org/10.1016/S0016-7037(98)00279–8

Azmy, K., Brand, U., Sylvester, P., Gleeson, S. A., Logan, A., Bitner, M. A., 2011. Biogenic and abiogenic low-Mg calcite (bLMC and aLMC): Evaluation of seawater-REE composition, water masses and carbonate diagenesis. *Chem. Geol.* 280, 180–90. https://doi.org/10.1016/j.chemgeo.2010.11.007

Banner, J. L., Hanson, G. N., Meyers, W. J., 1988. Rare earth element and Nd isotopic variations in regionally extensive dolomites from the Burlington-Keokuk Formation (Mississippian): Implications for REE mobility during carbonate diagenesis. *J. Sediment. Res.* 58, 415–32.

Bau, M., 1999. Scavenging of dissolved yttrium and rare earths by precipitating iron oxyhydroxide: Experimental evidence for Ce oxidation, Y-Ho fractionation, and lanthanide tetrad effect. *Geochim. Cosmochim. Acta* 63, 67–77.

Bau, M., Dulski, P., 1999. Comparing yttrium and rare earths in hydrothermal fluids from the Mid-Atlantic Ridge: Implications for Y and REE behaviour during near-vent mixing and for the Y/Ho ratio of Proterozoic seawater. *Chem. Geol.* 155, 77–90.

Bau, M., Koschinsky, A., 2009. Oxidative scavenging of cerium on hydrous Fe oxide: Evidence from the distribution of rare earth elements and yttrium between Fe oxides and Mn oxides in hydrogenetic ferromanganese crusts. *Geochem. J.* 43, 37–47.

Bau, M., Möller, P., Dulski, P., 1997. Yttrium and lanthanides in eastern Mediterranean seawater and their fractionation during redox-cycling. *Mar. Chem.* 56, 123–31. https://doi.org/10.1016/S0304-4203(96)00091–6

Bau, M., Schmidt, K., Koschinsky, A., Hein, J., Kuhn, T., Usui, A., 2014. Discriminating between different genetic types of marine ferro-manganese crusts and nodules based on rare earth elements and yttrium. *Chem. Geol.* 381, 1–9.

Bratina, B. J., Stevenson, B. S., Green, W. J., Schmidt, T. M., 1998. Manganese reduction by microbes from oxic regions of the Lake Vanda (Antarctica) water column. *Appl. Environ. Microbiol.* 64, 3791–7.

Byrne, R. H., Kim, K.-H., 1990. Rare earth element scavenging in seawater. *Geochim. Cosmochim. Acta* 54, 2645–56. https://doi.org/10.1016/0016–7037(90)90002–3

Cao, C., Liu, X.-M., Bataille, C. P., Liu, C., 2020. What do Ce anomalies in marine carbonates really mean? A perspective from leaching experiments. *Chem. Geol.* 532, 119413. https://doi.org/10.1016/j.chemgeo.2019.119413

Clarkson, M. O., Poulton, S. W., Guilbaud, R., Wood, R. A., 2014. Assessing the utility of Fe/Al and Fe-speciation to record water column redox conditions in carbonate-rich sediments. *Chem. Geol.* 382, 111–22.

Clement, B. G., Luther III, G. W., Tebo, B. M., 2009. Rapid, oxygen-dependent microbial Mn(II) oxidation kinetics at sub-micromolar oxygen concentrations in the Black Sea suboxic zone. *Geochim. Cosmochim. Acta* 73, 1878–89. https://doi.org/10.1016/j.gca.2008.12.023

Daye, M., Klepac-Ceraj, V., Pajusalu, M., Rowland, S., Farrell-Sherman, A., Beukes, N., Tamura, N., Fournier, G., Bosak, T., 2019. Light-driven anaerobic microbial oxidation of manganese. *Nature* 576, 311–14. https://doi.org/10.1038/s41586-019–1804-0

De Baar, H. J. W., German, C. R., Elderfield, H., Van Gaans, P., 1988. Rare earth element distributions in anoxic waters of the Cariaco Trench. *Geochim. Cosmochim. Acta* 52, 1203–19. https://doi.org/10.1016/0016–7037(88)90275-X

De Carlo, E. H., 2000. Rare earth element fractionation in hydrogenetic Fe-Mn crusts: The influence of carbonate complexation and phosphatization on Sm/Yb ratios. *Soc. Sediment. Geol.* 66, 271–85.

De Carlo, E. H., Green, W. J., 2002. Rare earth elements in the water column of Lake Vanda, McMurdo Dry Valleys, Antarctica. *Geochim. Cosmochim. Acta* 66, 1323–33. https://doi.org/10.1016/S0016-7037(01)00861–4

Edmonds, H. N., German, C. R., 2004. Particle geochemistry in the rainbow hydrothermal plume, Mid-Atlantic Ridge. *Geochim. Cosmochim. Acta* 68, 759–72.

Fowler, S., Hamilton, T., Peinert, R., La Rosa, J., Teyssie, J., 1992. The vertical flux of rare earth elements in the northwestern Mediterranean. In J. M. Martin and H. Barth, eds., *EROS 2000 (European River Ocean System): Proceedings of the Third Workshop on the North-West Mediterranean Sea*. Water Pollution Research Reports, 28. Brussels: Commission of the European Communities, 401–12.

German, C. R., Elderfield, H., 1990a. Application of the Ce anomaly as a paleoredox indicator: The ground rules. *Paleoceanography* 5, 823–33. https://doi.org/199010.1029/PA005i005p00823

German, C. R., Elderfield, H., 1990b. Rare earth elements in the NW Indian Ocean. *Geochim. Cosmochim. Acta* 54, 1929–40. https://doi.org/10.1016/0016–7037(90)90262-J

German, C. R., Elderfield, H., 1989. Rare earth elements in Saanich Inlet, British Columbia, a seasonally anoxic basin. *Geochim. Cosmochim. Acta* 53, 2561–71. https://doi.org/10.1016/0016–7037(89)90128–2

German, C. R., Holliday, B. P., Elderfield, H., 1991. Redox cycling of rare earth elements in the suboxic zone of the Black Sea. *Geochim. Cosmochim. Acta* 55, 3553–8. https://doi.org/10.1016/0016–7037(91)90055-A

Haley, B. A., Klinkhammer, G. P., McManus, J., 2004. Rare earth elements in pore waters of marine sediments. *Geochim. Cosmochim. Acta* 68, 1265–79. https://doi.org/10.1016/j.gca.2003.09.012

Haley, B. A., Klinkhammer, G. P., Mix, A. C., 2005. Revisiting the rare earth elements in foraminiferal tests. *Earth Planet. Sci. Lett.* 239, 79–97. https://doi .org/10.1016/j.epsl.2005.08.014

Hood, A. vS., Planavsky, N. J., Wallace, M. W., Wang, X. , 2018. The effects of diagenesis on geochemical paleoredox proxies in sedimentary carbonates. *Geochim. Cosmochim. Acta* 232, 265–87. https://doi.org/10.1016/j .gca.2018.04.022

Hood, A. vS., Wallace, M. W., 2014. Marine cements reveal the structure of an anoxic, ferruginous Neoproterozoic ocean. *J. Geol. Soc.* 171(6), 741–44.

Jacobs, L., Emerson, S., Huested, S. S., 1987. Trace metal geochemistry in the Cariaco Trench. *Deep Sea Res. Part Oceanogr. Res. Pap.* 34, 965–81.

Johnson, K. S., Berelson, W. M., Coale, K. H., Coley, T. L., Elrod, V. A., Fairey, W. R., Iams, H. D., Kilgore, T. E., Nowicki, J. L., 1992. Manganese flux from continental margin sediments in a transect through the oxygen minimum. *Science* 257, 1242–5. https://doi.org/10.1126/science.257.5074.1242

Kato, Y., Kano, T., Kunugiza, K., 2002. Negative Ce anomaly in the Indian banded iron formations: Evidence for the emergence of oxygenated deep-sea at 2.9–2.7 Ga. *Resour. Geol.* 52, 101–10. https://doi.org/10.1111/j.1751–3928 .2002.tb00123.x

Kato, Y., Ohta, I., Tsunematsu, T., Watanabe, Y., Isozaki, Y., Maruyama, S., Imai, N., 1998. Rare earth element variations in mid-Archean banded iron formations: Implications for the chemistry of ocean and continent and plate tectonics. *Geochim. Cosmochim. Acta* 62, 3475–97. https://doi.org/10.1016 /S0016-7037(98)00253–1

Kim, J.-H., Torres, M. E., Haley, B. A., Kastner, M., Pohlman, J. W., Riedel, M., Lee, Y.-J., 2012. The effect of diagenesis and fluid migration on rare earth element distribution in pore fluids of the northern Cascadia accretionary margin. *Chem. Geol.* 291, 152–65. https://doi.org/10.1016/j.chemgeo.2011.10.010

Klinkhammer, G. P., Bender, M. L., 1980. The distribution of manganese in the Pacific Ocean. *Earth Planet. Sci. Lett.* 46, 361–84. https://doi.org/10.1016 /0012-821X(80)90051–5

Koeppenkastrop, D., De Carlo, E. H., 1992. Sorption of rare-earth elements from seawater onto synthetic mineral particles: An experimental approach. *Chem. Geol.* 95, 251–63.

Koeppenkastrop, D., De Carlo, E. H., 1993. Uptake of rare earth elements from solution by metal oxides. *Environ. Sci. Technol.* 27, 1796–1802.

Koeppenkastrop, D., De Carlo, E., Roth, M., 1991. A method to investigate the interaction of rare earth elements in aqueous solution with metal oxides. *J. Radioanal. Nucl. Chem.* 152, 337–46.

Lawrence, M. G., Greig, A., Collerson, K. D., Kamber, B. S., 2006. Rare earth element and yttrium variability in south east Queensland waterways. *Aquat. Geochem.* 12, 39–72. https://doi.org/10.1007/s10498-005-4471-8

Lawrence, M. G., Kamber, B. S., 2006. The behaviour of the rare earth elements during estuarine mixing – revisited. *Mar. Chem.* 100, 147–61. https://doi.org /10.1016/j.marchem.2005.11.007

Lenton, T. M., Boyle, R. A., Poulton, S. W., Shields-Zhou, G. A., Butterfield, N. J., 2014. Co-evolution of eukaryotes and ocean oxygenation in the Neoproterozoic era. *Nat. Geosci.* 7.4, 257–65.

Lewis, B., Landing, W., 1991. The biogeochemistry of manganese and iron in the Black Sea. *Deep Sea Res. Part Oceanogr. Res. Pap.* 38, S773–S803.

Li, F., Webb, G. E., Algeo, T. J., Kershaw, S., Lu, C., Oehlert, A. M., Gong, Q., Pourmand, A., Tan, X., 2019. Modern carbonate ooids preserve ambient aqueous REE signatures. *Chem. Geol.* 509, 163–77. https://doi.org/10.1016/j .chemgeo.2019.01.015

Liu, X.-M., Hardisty, D. S., Lyons, T. W., Swart, P. K., 2019. Evaluating the fidelity of the cerium paleoredox tracer during variable carbonate diagenesis on the Great Bahamas Bank. *Geochim. Cosmochim. Acta* 248, 25–42. https:// doi.org/10.1016/j.gca.2018.12.028

Macleod, K. G., Irving, A. J., 1996. Correlation of cerium anomalies with indicators of paleoenvironment. *J. Sediment. Res.* 66, 948–55.

Mitra, A., Elderfield, H., Greaves, M., 1994. Rare earth elements in submarine hydrothermal fluids and plumes from the Mid-Atlantic Ridge. *Mar. Chem.* 46, 217–35.

Moffett, J. W., 1990. Microbially mediated cerium oxidation in sea water. *Nature* 345, 421–3. https://doi.org/10.1038/345421a0

Möller, P., Bau, M., 1993. Rare-earth patterns with positive cerium anomaly in alkaline waters from Lake Van, Turkey. *Earth Planet. Sci. Lett.* 117, 671–6. https://doi.org/10.1016/0012-821X(93)90110-U

Nothdurft, L. D., Webb, G. E., Kamber, B. S., 2004. Rare earth element geochemistry of Late Devonian reefal carbonates, Canning Basin, Western Australia: Confirmation of a seawater REE proxy in ancient limestones.

Geochim. Cosmochim. Acta 68, 263–83. https://doi.org/10.1016/S0016-7037(03)00422-8

O'Connell, B., Wallace, M. W., Hood, A. vS., Lechte, M. A., Planavsky, N. J., 2020. Iron-rich carbonate tidal deposits, Angepena Formation, South Australia: A redox-stratified Cryogenian basin. *Precambrian Res.* 342, 105668. https://doi.org/10.1016/j.precamres.2020.105668

Ohta, A., Kawabe, I., 2001. REE (III) adsorption onto Mn dioxide (MnO_2) and Fe oxyhydroxide: Ce (III) oxidation by MnO_2. *Geochim. Cosmochim. Acta* 65, 695–703.

Planavsky, N., Bekker, A., Rouxel, O. J., Kamber, B., Hofmann, A., Knudsen, A., Lyons, T. W., 2010. Rare earth element and yttrium compositions of Archean and Paleoproterozoic Fe formations revisited: New perspectives on the significance and mechanisms of deposition. *Geochim. Cosmochim. Acta* 74, 6387–6405. https://doi.org/10.1016/j.gca.2010.07.021

Pourmand, A., Dauphas, N., Ireland, T. J., 2012. A novel extraction chromatography and MC-ICP-MS technique for rapid analysis of REE, Sc and Y: Revising CI-chondrite and post-Archean Australian Shale (PAAS) abundances. *Chem. Geol.* 291, 38–54. https://doi.org/10.1016/j.chemgeo.2011.08.011

Quinn, K. A., Byrne, R. H., Schijf, J., 2006. Sorption of yttrium and rare earth elements by amorphous ferric hydroxide: Influence of solution complexation with carbonate. *Geochim. Cosmochim. Acta* 70, 4151–65. https://doi.org/10.1016/j.gca.2006.06.014

Saager, P. M., De Baar, H. J. W., Burkill, P. H., 1989. Manganese and iron in Indian Ocean waters. *Geochim. Cosmochim. Acta* 53, 2259–67. https://doi.org/10.1016/0016-7037(89)90348-7

Saager, P. M., Schijf, J., De Baar, H. J. W., 1993. Trace-metal distributions in seawater and anoxic brines in the eastern Mediterranean Sea. *Geochim. Cosmochim. Acta* 57, 1419–32. https://doi.org/10.1016/0016-7037(93)90003-F

Sherrell, R. M., Field, M. P., Ravizza, G., 1999. Uptake and fractionation of rare earth elements on hydrothermal plume particles at 9 45°N, East Pacific Rise. *Geochim. Cosmochim. Acta* 63, 1709–22.

Shields, G., Stille, P., 2001. Diagenetic constraints on the use of cerium anomalies as palaeoseawater redox proxies: An isotopic and REE study of Cambrian phosphorites. *Chem. Geol.* 175, 29–48. https://doi.org/10.1016/S0009-2541(00)00362-4

Shields, G., Webb, G., 2004. Has the REE composition of seawater changed over geological time? *Chem. Geol.* 204.1, 103–7.

Sholkovitz, E. R., Landing, W. M., Lewis, B. L., 1994. Ocean particle chemistry: The fractionation of rare earth elements between suspended particles and seawater. *Geochim. Cosmochim. Acta* 58, 1567–79. https://doi.org/10.1016 /0016-7037(94)90559-2

Sunda, W. G., Huntsman, S. A., 1988. Effect of sunlight on redox cycles of manganese in the southwestern Sargasso Sea. *Deep Sea Res. Part Oceanogr. Res. Pap.* 35, 1297–1317.

Tang, D., Shi, X., Wang, X., Jiang, G., 2016. Extremely low oxygen concentration in mid-Proterozoic shallow seawaters. *Precambrian Res.* 276, 145–57. https://doi.org/10.1016/j.precamres.2016.02.005

Tostevin, R., Mills, B. J. W., 2020. Reconciling proxy records and models of Earth's oxygenation during the Neoproterozoic and Palaeozoic. *Interface Focus* 10, 20190137. https://doi.org/10.1098/rsfs.2019.0137

Tostevin, R., Shields, G. A., Tarbuck, G. M., He, T., Clarkson, M. O., Wood, R. A., 2016a. Effective use of cerium anomalies as a redox proxy in carbonate-dominated marine settings. *Chem. Geol.* 438, 146–62. https://doi .org/10.1016/j.chemgeo.2016.06.027

Tostevin, R., Wood, R. A., Shields, G. A., Poulton, S. W., Guilbaud, R., Bowyer, F., Penny, A. M., He, T., Curtis, A., Hoffmann, K. H., Clarkson, M. O., 2016b. Low-oxygen waters limited habitable space for early animals. *Nat. Commun.* 7. https://doi.org/10.1038/ncomms12818

Trefry, J. H., Presley, B. J., Keeney-Kennicutt, W. L., Trocine, R. P., 1984. Distribution and chemistry of manganese, iron, and suspended particulates in Orca Basin. *Geo-Mar. Lett.* 4, 125–30. https://doi.org/10.1007/BF02277083

Wallace, M. W., Hood, A. vS., Shuster, A., Greig, A., Planavsky, N. J., Reed, C. P. , 2017. Oxygenation history of the Neoproterozoic to early Phanerozoic and the rise of land plants. *Earth Planet. Sci. Lett.* 466, 12–19. https://doi.org/10.1016/j.epsl.2017.02.046

Webb, G. E., Kamber, B. S., 2000. Rare earth elements in Holocene reefal microbialites: A new shallow seawater proxy. *Geochim. Cosmochim. Acta* 64, 1557–65. https://doi.org/10.1016/S0016-7037(99)00400-7

Webb, G. E., Nothdurft, L. D., Kamber, B. S., Kloprogge, J. T., Zhao, J.-X., 2009. Rare earth element geochemistry of scleractinian coral skeleton during meteoric diagenesis: A sequence through neomorphism of aragonite to calcite. *Sedimentology* 56, 1433–63. https://doi.org/10.1111/j.1365-3091.2008.01041.x

Zaky, A. H., Brand, U., Azmy, K., Logan, A., Hooper, R. G., Svavarsson, J., 2016. Rare earth elements of shallow-water articulated brachiopods: A bathymetric sensor. *Palaeogeogr. Palaeoclimatol. Palaeoecol.* 461, 178–94. https://doi.org/10.1016/j.palaeo.2016.08.021

Zhang, K., Zhu, X.-K., Yan, B., 2015. A refined dissolution method for rare earth element studies of bulk carbonate rocks. *Chem. Geol.* 412, 82–91. https://doi.org/10.1016/j.chemgeo.2015.07.027

Cambridge Elements ≡

Elements in Geochemical Tracers in Earth System Science

Timothy Lyons
University of California

Timothy Lyons is a Distinguished Professor of Biogeochemistry in the Department of Earth Sciences at the University of California, Riverside. He is an expert in the use of geochemical tracers for applications in astrobiology, geobiology and Earth history. Professor Lyons leads the 'Alternative Earths' team of the NASA Astrobiology Institute and the Alternative Earths Astrobiology Center at UC Riverside.

Alexandra Turchyn
University of Cambridge

Alexandra Turchyn is a University Reader in Biogeochemistry in the Department of Earth Sciences at the University of Cambridge. Her primary research interests are in isotope geochemistry and the application of geochemistry to interrogate modern and past environments.

Chris Reinhard
Georgia Institute of Technology

Chris Reinhard is an Assistant Professor in the Department of Earth and Atmospheric Sciences at the Georgia Institute of Technology. His research focuses on biogeochemistry and paleoclimatology, and he is an Institutional PI on the 'Alternative Earths' team of the NASA Astrobiology Institute.

About the Series

This innovative series provides authoritative, concise overviews of the many novel isotope and elemental systems that can be used as 'proxies' or 'geochemical tracers' to reconstruct past environments over thousands to millions to billions of years – from the evolving chemistry of the atmosphere and oceans to their cause-and-effect relationships with life.

Covering a wide variety of geochemical tracers, the series reviews each method in terms of the geochemical underpinnings, the promises and pitfalls, and the 'state-of-the-art' and future prospects, providing a dynamic reference resource for graduate students, researchers and scientists in geochemistry, astrobiology, paleontology, paleoceanography and paleoclimatology.

The short, timely, broadly accessible papers provide much-needed primers for a wide audience – highlighting the cutting-edge of both new and established proxies as applied to diverse questions about Earth system evolution over wide-ranging time scales.

Cambridge Elements $\equiv$

Elements in Geochemical Tracers in Earth System Science

Elements in the Series

A full series listing is available at: www.cambridge.org/EESS